The Nature of the Caribbean

Alfonso Silva Lee

Illustrated by

Alexis Lago

PANGAEA

Saint Paul

ISBN 1929165153 Hardcover
ISBN 1929165145 Paperback
ISBN 1929165161 eBook

Library of Congress CIP Data

Silva Lee, Alfonso.
My island and I : the nature of the
Caribbean / Alfonso Silva Lee ;
p. cm.
ISBN 1-929165-15-3 (hardcover)--
ISBN 1-929165-14-5 (paperback)--
ISBN 1-929165-16-1 (eBook)
1. Island ecology--Caribbean Area--Juvenile
literature. [1. Island ecology--Caribbean Area.
2. Ecology.] I. Title.
QH109.A1 S56 2001
577.5'2'09729--dc21

2001001567

Published in the
United States of America by

P A N G A E A
226 Wheeler Street South
Saint Paul, MN 55105-1927 USA
Tel 651.690-3320 Fax 651.690-1485
www.pangaea.org info@pangaea.org

First Edition

2001

To Gabriela, Mercedes, Virginia, Juan José
and Little Raymond, and to the lizards,
all of whom liven up the islands.

Islands in Good Company

The Caribbean islands are in the middle of the sea, yet they are not alone. The islands and the sea are very good friends: every day they give each other water.

Puerto Rico, Martinique, Cuba, Bonaire, the Virgin Islands and dozens of others deliver water from their rivers to the ocean. Manatees live in this rather sweet water, along with fish, crabs and shrimp who breed there.

The sun is everybody's friend, and also the sea's. Every day the sun warms up the sea. Then, little by little, the water mixes with the air and clouds are formed. When the clouds get big, it rains a lot. That's how the sea offers water to the islands. This is the water we drink, the water we bathe with and the water that makes plants grow.

Taínos

The first aborigines arrived in the islands some 6,000 years ago in very long canoes. Their grandparents' grandparents had come to the Caribbean from the South American continent.

They were the first people to fall in love with these islands, where there still was not a single house, nor a road or a bridge.

The aborigines must have been very happy to arrive on these islands. They built palm huts and canoes, grew plants needed for food and wove their own fishing nets.

With time, they came to love the mountains and the rivers, and the fresh breeze. Each island turned into their big home, and they changed more and more, until they became part of the islands, too. They became Tainos.

Very, Very Old Islands

The islands, though, existed for quite a long time before the aborigines—about 35 million years. That is a great amount of time. Can you guess how long a line of 35 million ants would be?

(A line of 35 million ants would be 160 kilometers/100 miles long!)

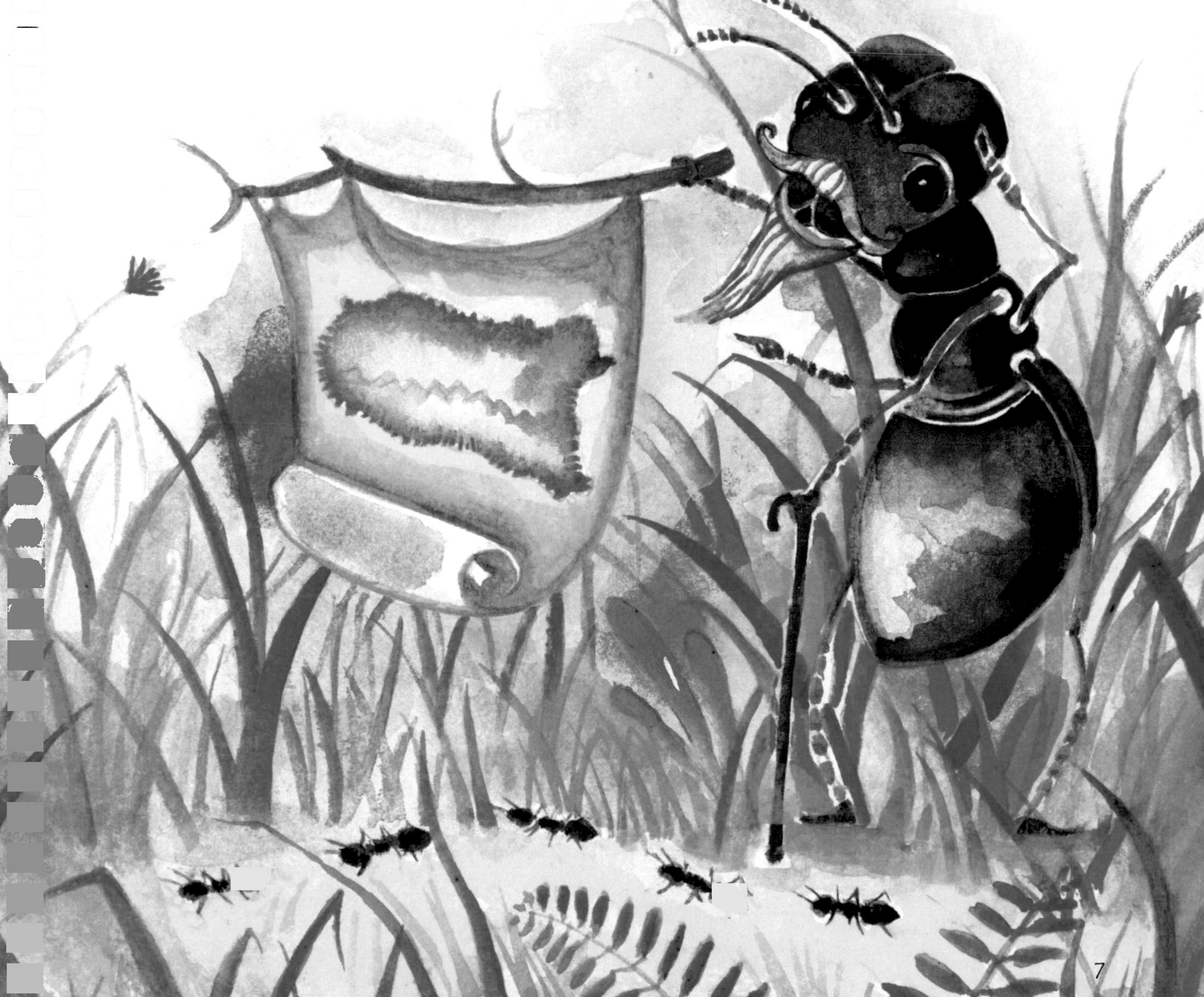

The Caribbean islands were born from under the sea. At the beginning they were very small, and on them not a single palm tree grew, not a lizard dashed or frog called.

But on the continents nearby, there were gigantic forests, crowded with insects, lizards, frogs, snakes and birds. It was these forests that gave the islands the grand-grandparents of the grand-grandparents of today's animals and plants.

In that distant time there were no post offices, ships, airplanes nor people, so each plant and animal made the trip on its own.

Traveling Plants

A trumpet tree and a fern spend their entire lives in the same place they were born. Trees and vines certainly do not seem to be travelers at all ... but they are! In their own way, plants are very intelligent. Although incapable of taking a single step, they manage to run, swim and even fly.

Their trick is in the seed. The seeds of some plants, for example, float. The creeks carry them to the rivers, and the rivers take them to the sea.

Once at sea, the currents push the seeds until they are thrown onto a beach days later. (Whenever you go to a Caribbean beach, check the shore. You may find seeds that have floated all the way from Venezuela or from Brazil.)

Other plants give very small seeds that are covered with thin hairs or that have wings. Strong winds carry these seeds very far, even from one island to another.

In order to travel, there are plants that make use of birds. Some have very small seeds that, along with mud, stick to the birds' feet. Other seeds are covered with a sticky substance or with little hooks. In this way they attach themselves to bird feathers and travel long distances.

And finally, there are plants—like guava—with seeds wrapped inside their fruit. These are eaten by some birds, which then carry the seeds in their bellies as they fly over water.

Traveling Animals

The grand-grandparents of the grand-grandparents of the islands' bees, butterflies and birds simply flew to the islands.

But frogs don't have wings... nor can they sit for long in saltwater. How do you think that the grand-grandparents of the grand-grandparents of the frogs could have traveled to the Caribbean islands?

Well, the little frogs made the trip like big-time sailors. Whenever hurricanes swept the South and Central American forests, the rivers overflowed and many animals fell into the torrents. To save their lives, frogs, lizards and millipedes climbed onto the branches carried by the water. The branches later floated out to sea and ocean currents pushed them towards new island homes.

The locust trees, cacti, butterflies, frogs, lizards and birds that live in the Caribbean islands today, therefore, are the grand-grandchildren of the grand-grandchildren of plants and animals that lived in the great continental forests of South and Central America.

Really Antillean

The Caribbean islands today form groups now called the Greater and Lesser Antilles. Their forests are rich. There are so many different palms and orchids! There are yellow, white and orange butterflies, and also others striped, or spotted like a leopard. There are many more plants and animals in the forest than in any city park.

The dog and cat, just like the horse, sheep and rooster, are not Antillean animals. Humans also brought the rose bush and the flamboyant tree to the islands from distant lands.

The broad-winged hawk, manatee and ceiba tree are,

indeed, Antillean. The broad-winged hawk and the seeds of the ceiba are excellent flyers, and the manatee a good swimmer. Like many other plants and animals, they travel frequently across the sea and can be found in Mexico, Colombia, St. Vincent, Guyana and Jamaica.

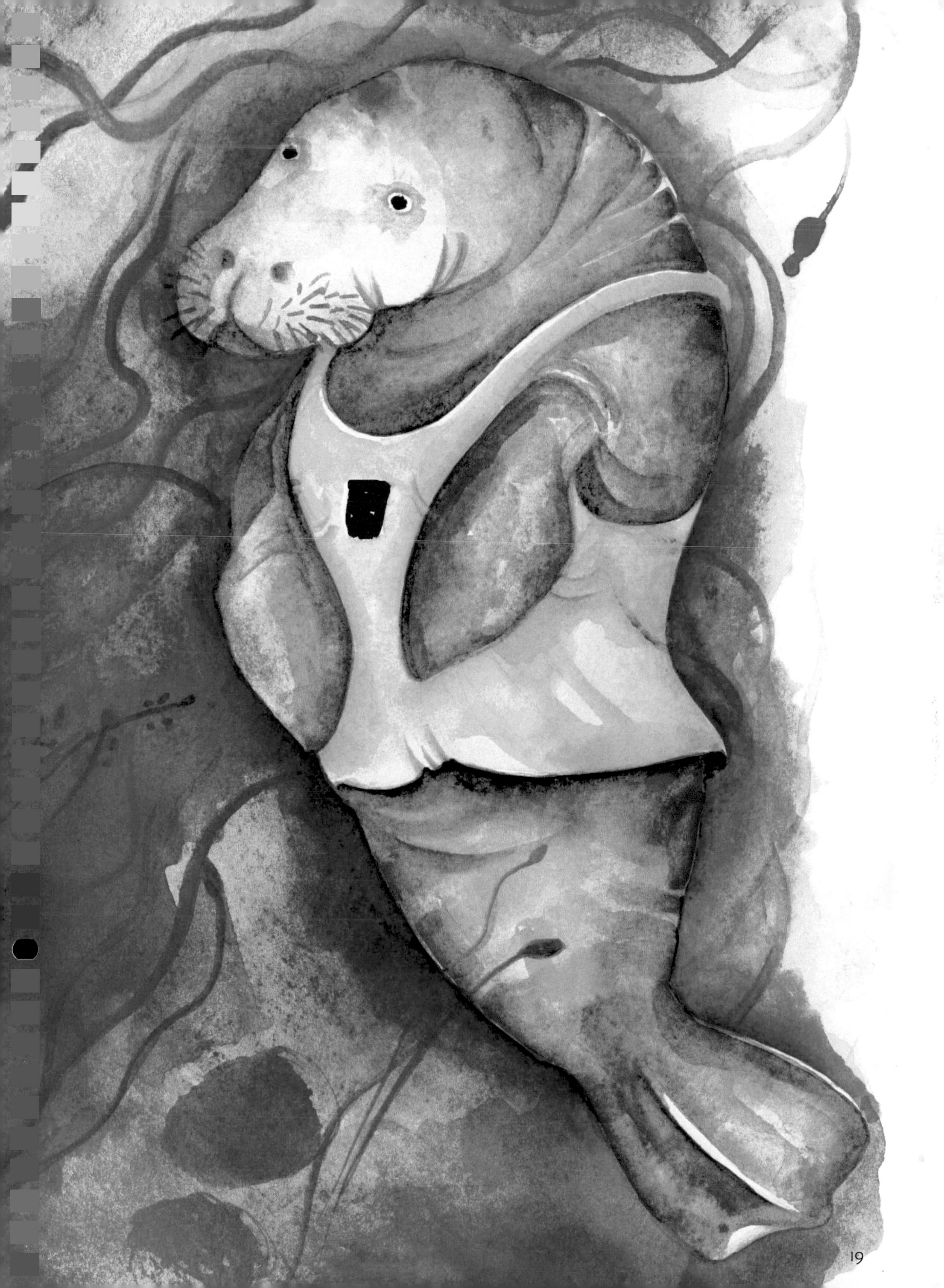

The Puerto Rican coquí frog, the Cuban boa snake, Jamaica's tody bird and Hispaniola's green ebony tree, however, live only on their own islands. Their ancestors arrived from the continent millions of years ago.

Since these creatures were incapable of flying or swimming long distances, little by little they adapted more and more to their own island's environment: they became unique and very special. Today they do not live in any other place in the world.

For that reason, the lizards you see during the day and the frogs you hear at night on each island are not only different from those that live in the rest of the American lands, but also different from those of neighboring Caribbean islands.

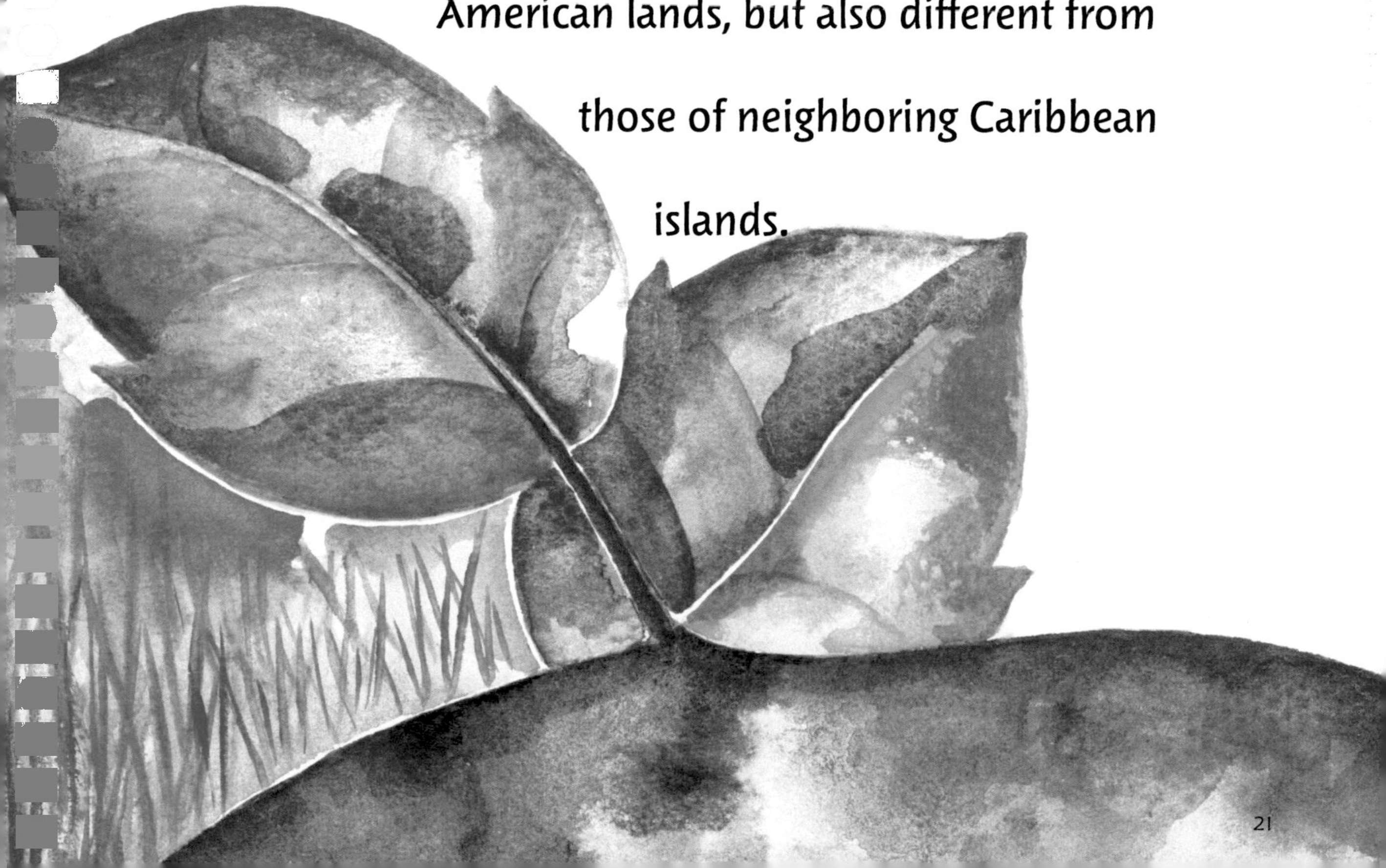

The Forest's Magic

The forest takes care of itself. We never have to send trucks into the forest to bring in food or take out garbage. The forest is so wonderful that is does not produce garbage at all. Everything is used; everything is recycled. The trees and other plants actually make the forest. Their magic is in their roots and leaves.

The roots take water and minerals from the soil. In order to do a better job, they are assisted by fungi, which include mushrooms. If the fungi or plants are on their own, they grow little, but when they live together they help each other grow strong and healthy.

Leaves are marvelous suntraps. Each leaf has millions of small particles that catch sunlight. These particles are the ones that give plants their green color.

Let's Swap Vapors!

In order to do their magic, the leaf uses a gas that is very abundant in the air. It has a long and hard-to-remember name: carbon dioxide. Luckily, it is also known as CO_2 (and easily pronounced, "see-o-two").

Using CO_2, water, minerals and sunlight, the leaves make food for the entire plant. In the process of building its own food, the plant releases another gas, called oxygen.

It is thanks to the plant's gift of oxygen that the muscles and brains of animals can function. While we walk, run and think, we convert the oxygen into CO_2 and release it from our lungs. Between the plants and the animals, then, there is also a very great friendship. Each one gives the other the gas it needs most!

Trees Turned into Fungi

With time, every tree turns old and dies just like every other living creature. Then a strong wind knocks it to the ground. Since the tree surely bore thousands of seeds and many offspring during its long life, this should not be cause for sadness. For many creatures, on the other hand, the occasion is a feast.

The first to show up at a fallen tree trunk are the fungi and bacteria. They are so small we cannot see them with our eyes. That is why their arrival is silent and invisible.

Without fungi and bacteria, wood would never rot. Little by little they both penetrate the whole trunk and feed on it until none of the tree remains. In this way the trunk is transformed into fungi and bacteria.

Trees Turned into Termites

Other fallen trees are eaten by termites. Termites look a lot like ants, but have a soft body and are whitish. Since termites live in complete darkness, they are blind. They find their way through the dark tunnels they make in the trees by means of smell and touch.

Wood is not only hard to cut, but also hard to digest. That's why we don't make sawdust hamburgers.

In order to cut wood, termites have very sharp and powerful "mandibles"—their jaws.

To be able to digest wood, termites have friendly bacteria in their stomachs. You and I also have friendly bacteria in our intestines—although different from those of termites—that help us digest meat and vegetables.

What Am I Made of?

We humans feed on many different organisms. We eat fungi, roots, leaves, flowers, fruits, seeds and even whole plants.

We also feed on land crabs and conchs, fish, birds,

pigs and cattle. In some countries our diet includes insects, frogs, iguanas, snakes, turtles, crocodiles, bats and even whales! Eating these organisms, we transform them into our own selves.

A bull, for example, is made of the grasses on which it feeds. The grass, on the other hand, is made of the minerals found in the soil, the rain brought by the clouds, and the light shining from the sun.

You and I eat hamburgers. But the hamburger comes from a bull. We are, therefore, made of bulls and also of grasses and mountains, of seawater, clouds and sun. This is quite magical.

From Soil to Red-tailed Hawk

There are snails, millipedes, ants, beetles and other insects that feed on fungi. By feeding on fungi, these animals transform fungi into themselves. But fungi are made of fallen tree trunks, as we have already seen. Snails, millipedes and insects are thus made not only of fungi, but also of trees.

The jumble is delicious, since the larger animals, by eating the smaller ones, transform these into themselves: termite becomes frog, which turns into green lizard, which transforms into red-tailed hawk…

When each creature dies, bacteria and fungi decompose it. This way the dead are again transformed into living organisms.

Therefore:

- the soil, clouds, air and sun are transformed into trees;
- the trees are transformed into bacteria and fungi;
- the bacteria and fungi are transformed into ants and snails;
- the ants and snails are transformed into lizards and frogs;

- the lizards and frogs are transformed into snakes and hawks;
- and the snakes and hawks are transformed into bacteria, fungi and soil—that is, into food for trees, ants and snails!

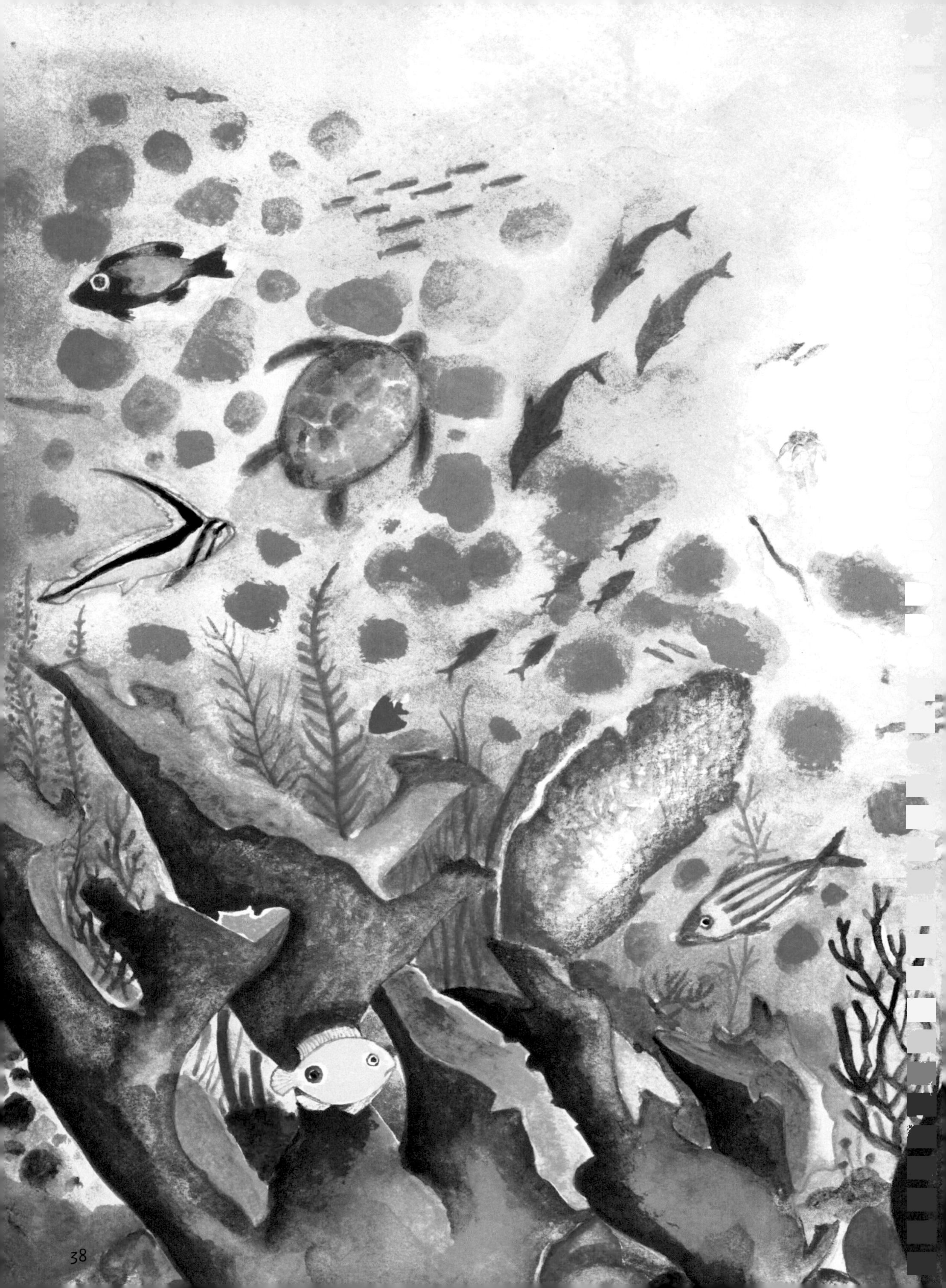

Algae and Corals

At sea everything is quite different from what is found in the forest. At the bottom of the ocean there is no soil—just sand or silt—and the sunlight never reaches the deeper parts.

But in the seas as well as in the forest, many different animals live, and also algae. There are abundant fish, crabs and snails, and also turtles and dolphins. Most algae are microscopic. Even though they cannot be seen with the naked eye, there are thousands and thousands of them in a cup of seawater.

Friendship among marine organisms is even greater than among the land ones. Underwater some creatures are transformed into others, which later are transformed into still others, which ..

The seawater that surrounds the islands is warm. That's why there are many corals, and reefs are formed there.

Corals don't look like any land animal. Each coral is made up of hundreds of polyps. And each polyp has tentacles: it looks like a newborn octopus placed upside down.

Coral polyps cannot move a single inch. They spend their entire lives in shelters they made themselves that are hard as rock and crowded with spines and sharp edges. With their tentacles wide open, the polyps wait for the arrival of tiny animals that are captured and swallowed.

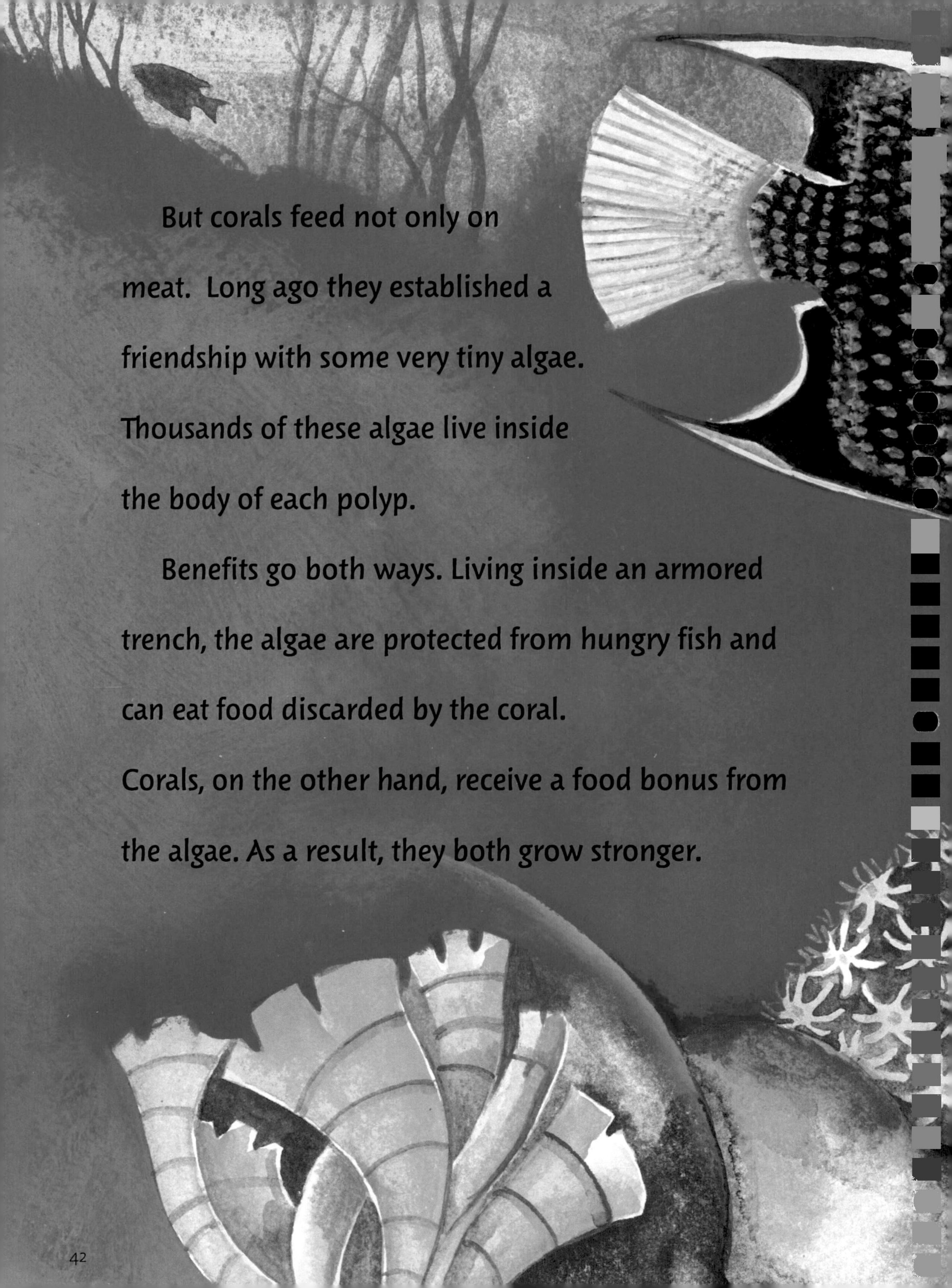

But corals feed not only on meat. Long ago they established a friendship with some very tiny algae. Thousands of these algae live inside the body of each polyp.

Benefits go both ways. Living inside an armored trench, the algae are protected from hungry fish and can eat food discarded by the coral. Corals, on the other hand, receive a food bonus from the algae. As a result, they both grow stronger.

Underwater Friendship

Tiny striped fishes the size of a match live in the coral reefs of the Greater and Lesser Antilles that are friends of all other fishes—friends even of the large-mouthed groupers and sharp-toothed barracuda.

The little fishes are called cleaning gobies, and they clean off the parasites attached to the skin of other

fishes—one by one. Parasites are their daily food.

When a fish living in a coral reef feels an itching on its skin, it swims to where the cleaning gobies are. Then a goby "lands" on the itchy fish and inspects every patch of skin. In search of food, the cleaning goby even goes into the fish's mouth and eats up every parasite.

Thanks to this friendship, the goby finds its favorite food, while the larger fish is cured of itching.

What Does a Lizard Think?

An Antillean lizard differs from us by being much smaller, having a very long tail and lacking any hair. But, on the other hand, we both have much in common.

The bones of its skeleton, for example, are similar to our own. It, too, has lungs for breathing, a heart, kidneys and a liver. In its head it has a pair of eyes similar to ours; on each side of its head it has ears. Inside its head, most importantly, it has a brain—quite small but complete.

If you watch a lizard for a while, you will be able to see it get interested in some little moth, dart off and capture it. The lizard, no doubt, was on the alert for flying insects, and has excellent sight. It also seems to find the moths as tasty as we regard chocolate ice cream.

If we get closer, the lizard will run and hide; it considers us, not unreasonably, to be dangerous giants. A while later we will see it search for a shady spot: a little heat is fine … but not too much!

Just like us, the lizard sleeps at night. It doesn't have a house or a bed, but as darkness arrives it searches for a leaf of its choosing, lays over it and closes its eyes until the next morning.

Who knows if, while making itself comfortable on the leaf—a bit tired perhaps from a busy day—it enjoys the reddish beauty of the setting sun! We would have to be a lizard to know that.

The Charm of Being Alive

It is marvelous to sense everything around us. How beautiful the clean blue sky, the sailing clouds and the pure-silver-like rain are! And what joy it is to feel surrounded by the water of a river or of the seashore!

Flowers, with all their thousand fragrances, are also pleasing and so are the sun's warmth, a landscape of very green mountains, contact with an animal friend and the touch of mom and dad.

All this takes place around you. No doubt about it, you are the center of the universe. But . . . wait a second . . . I am also the center of the universe. And so are your mother and father, and each of your friends, each person . . .

One Very Big Family

Each lizard is also the center of its own world, and so is each coral polyp, each grasshopper and each vine. All of them are also made of mountains and sunshine.

It is a great delight that each is the center of everything, and that each is made of all others. Neither you nor I can live without the others. We are made of them and they are made of us.

The Caribbean islands are mountains, rivers, beaches and lagoons. Guadeloupe, Aruba, Saba, the Grenadines and all the other Antillean islands are made of orquids and spiders, night-blooming cacti and frogs, todies and bats. You and I also belong to this very big family.

Acknowledgments

Big thanks to the forests and reefs of the Caribbean, to each tree, fungus, snail, coral, crab and fish. Also thanks, of course, to the mountains, the clouds, the sea, the rivers and sun.

Many friends provided nourishing criticism and/or support:

Sonia Aponte Zeno
Carmen I. Asencio
Coloma Araújo
Jacqueline Biscombe
Harry Caraballo Oliveras
Edwin Carrasquillo
Miguel García Bermúdez
Víctor L. González
Juan José González
José Roberto Martínez
Néstor Murray-Irizarry
Norma Padilla
Alberto Areces
Bonnie Hayskar
Brad Richter
Lauren Raz
María Sánchez
John Guarnaccia
F. Javier Saracho
Tania Serrallés
Gabriela Silva
Luis Alfonso Silva
Ádrianne Tossas

CHILDREN 4-8

Have you ever felt like an island?... You and the islands of the Caribbean are one and the same, along with the sea so blue, the sailing clouds, the whirling moon, the golden sun. Rainbows, giant trees, tiny tiny frogs and speedy lizards are not just neighbors, but real friends and real family.

Alexis and Alfonso are both Caribbean islanders. Here, they invite you to feel like an island.

1 929165 1
9 78